David Peck Todd

A Continuation of de Damoiseau's Tables of the Satellites of

Jupiter

David Peck Todd

A Continuation of de Damoiseau's Tables of the Satellites of Jupiter

ISBN/EAN: 9783337365516

Printed in Europe, USA, Canada, Australia, Japan

Cover: Foto ©berggeist007 / pixelio.de

More available books at **www.hansebooks.com**

A CONTINUATION

OF

DE DAMOISEAU'S TABLES

OF THE

SATELLITES OF JUPITER,

TO

THE YEAR 1900.

BY

D. P. TODD, B. A.

PUBLISHED FOR THE AMERICAN EPHEMERIS AND NAUTICAL ALMANAC,
BY AUTHORITY OF THE SECRETARY OF THE NAVY.

WASHINGTON:
BUREAU OF NAVIGATION.
1876.

PREFATORY NOTE.

THE *Tables Écliptiques des Satellites de Jupiter, d'après la Théorie de leurs attractions mutuelles et les constantes déduites des Observations,* par le Baron DE DAMOISEAU, Paris, 1836, terminate with the epoch 1880.0. Entirely new tables of the satellites, being very laborious to construct, have not yet been published. .In this continuation of the *Tables Écliptiques,* no changes in the fundamental formulæ and elements have been made, it being believed that the consequent inconvenience to the future investigator of the motions of the satellites would more than neutralize any advantages supposed to arise from such a change.

The work was planned early in the present year: the definitive computations were not, however, commenced until August.

An acknowledgment of indebtedness is due Professor NEWCOMB, from whom advice has been received, from time to time, during the prosecution of the work.

D. P. TODD.

WASHINGTON, 1876, *November the 7th.*

CONTENTS.

INTRODUCTION.

THE CONSTRUCTION OF THE TABLES.

INTRODUCTION.

THE CONSTRUCTION OF THE TABLES.

THE fundamental formulae and elements of the *Tables Écliptiques* are given by DE DAMOISEAU, Introduction, pages (1)—(x).

THE TABLES OF THE ECLIPSES. — Of the Tables for computing the eclipses, those which require extension are I and III of each satellite.

THE FORMATION OF TABLE I.

The First Satellite.—Counting from the first mean conjunction in each year, and letting i represent the number of mean synodic revolutions, then

$$\text{from 1750 to 1880,} \qquad i = 26827;$$
$$\text{1880 to 1900,} \qquad i = 4128.$$

The epoch of the first mean conjunction in the year 1880 is January 1, 18^h 16^m $45^s.070$, Paris mean civil time.

$$\text{The mean synodic revolution of I} \times 206 = 361^d \ 14^h \ 11^m \ 24^s.74721$$
$$\times 207 = 366 \ \ 8 \ \ 40 \ \ 0.69259$$

The terms of the arguments of the inequalities which increase uniformly with the time are based on the data of the following table:

Argument.	Terms.	Adopted value for first mean conj. 1880.	Motion for $i = 206$.	Motion for $i = 207$.
		$^s.$ ^{o}	^{o}	^{o}
1	$u_o - \pi_o$	11 8.4926	30.2924	30.4395
2	$U - \pi'$	11 29.9085	359.3412	1.0856
3	$U - u_o \quad -7°.43$	9 12.4537	329.0501	330.6471
4	$u_1 - u_{II} \quad -2°.77$	4 23.9896	149.9562	330.6841
5	$u_1 - u_{III} \quad -4°.10$	4 6.0301	44.9342	316.0261
6	$u_1 - \pi_{III} \quad -5°.51$	1 20.2086	27.6958	27.8302
7	$u_1 - \pi_{IV} \quad -5°.51$	2 12.4348	29.5979	29.7416
8	$u_1 - 2u_{II} + \pi_{III}$	7 28.0372	272.2184	273.5399
9	$u_1 - 2u_{II} + \pi_{IV}$	7 5.5824	270.3145	271.6267
I	$u_1 - \Pi \quad -5°.51$	1 0.2171	30.2944	30.4415
II	$u_1 - \Lambda_{II} \quad -5°.51$	0 11.7738	42.3512	42.5568
III	$u_1 - \Lambda_{III} \quad -5°.51$	4 22.1023	32.8438	33.0032

The Second Satellite.—Counting from the first mean conjunction in each year,

$$\text{from 1750 to 1880,} \qquad i = 13360;$$
$$\text{1880 to 1900,} \qquad i = 2055.$$

The epoch of the first mean conjunction in the year 1880 is January 3, 5^h 3^m $2^s.719$, Paris mean civil time.

$$\text{The mean synodic revolution of II} \times 102 = 362^d \ 12^h \ 25^m \ 20^s.99381$$
$$\times 103 = 366 \ \ 1 \ \ 43 \ \ 14.72904$$

The terms of the arguments of the inequalities which increase uniformly with the time are based on the data of the following table:

Argument.	Terms.	Adopted value for first mean conj. 1880.	Motion for $i = 102$.	Motion for $i - 103$.
1	$u_o - \pi_o$	11ʳ 8.6129	30°.1201	30°.4151
2	$U - \pi'$	0 1.3362	357.2974	0.8003
3	$U - u_o \quad -7°.43$	9 13.7591	327.1786	330.3862
4	$u_{II} - u_{III} - 2°.78$	1 24.5665	149.1032	330.5650
5	$u_{II} - \pi_{III} - 5°.51$	1 20.0917	27.5365	27.8065
6	$u_{II} - \pi_{IV} - 5°.51$	2 12.5521	29.4296	29.7181
7	$u_I - 2u_{II} + \pi_{III}$	7 29.1189	270.6701	273.3238
8	$u_I - 2u_{II} + \pi_{IV}$	7 6.6516	268.7771	271.4121
I	$u_{II} - \Pi \quad -5°.51$	1 0.3335	30.1221	30.4174
II	$u_{II} - \Lambda_{II} - 5°.51$	0 11.9208	42.1103	42.5231
III	$u_{II} - \Lambda_{III} - 5°.51$	4 22.2343	32.6570	32.9772
IV	$u_{II} - \Lambda_{IV} - 5°.51$	5 5.3055	30.8085	31.1105

The Third Satellite.— Counting from the first mean conjunction in each year,

from 1750 to 1880, $i = 6625$;
1880 to 1900, $i = 1019$.

The epoch of the first mean conjunction in the year 1880 is January 1, 8ʰ 27ᵐ 5ˢ.256, Paris mean civil time.

The mean synodic revolution of III $\times$ 50 = 358ᵈ 7ʰ 39ᵐ 52ˢ.70985
$\times$ 51 = 365 11 39 26.56105
$\times$ 52 = 372 15 39 4.41825

The terms of the arguments of the inequalities which increase uniformly with the time are based on the data of the following table:

Argument.	Terms.	Adopted value for first mean conj. 1880.	Motion for $i = 50$.	Motion for $i = 51$.	Motion for $i = 52$.
1	$u_o - \pi_o$	11ʳ 8.7077	29°.7713	30°.3667	30°.9622
2	$U - \pi'$	0 2.4625	353.1596	0.2228	7.2860
3	$U - u_o \quad -7°.43$	9 14.7905	323.3896	329.8574	336.3252
4	$u_{II} - u_{III} - 5°.60$	3 20.0286	294.7529	300.6479	306.5430
5	$u_{III} - u_{IV} - 3°.20$	9 17.5541	220.4107	66.4196	272.4584
6	$u_{III} - \pi_{III} - 5°.51$	1 20.1813	27.2177	27.7620	28.3064
7	$u_{III} - \pi_{IV} - 5°.51$	2 12.6504	29.0888	29.6706	30.2523
8	$u_I - 2u_{II} + \pi_{III}$	7 29.9752	267.5356	272.8863	278.2370
9	$u_I - 2u_{II} + \pi_{IV}$	7 7.5018	265.6644	270.9777	276.2910
I	$u_{III} - \Pi \quad -5°.51$	1 0.3216	29.7724	30.3679	30.9633
II	$u_{III} - \Lambda_{II} - 5°.51$	4 22.3101	32.2768	32.9244	33.5700
III	$u_{III} - \Lambda_{IV} - 5°.51$	5 5.4064	30.4517	31.0608	31.6698
IV	$u_{III} - \Lambda_{II} - 5°.51$	0 11.9418	41.6217	42.4511	43.2865

The Fourth Satellite.—-Counting from the first mean conjunction in each year,

from 1750 to 1880, $i = 2834$;
1880 to 1900, $i = 436$.

The epoch of the first mean conjunction in the year 1880 is January 1, 3ʰ 6ᵐ 37ˢ.286, Paris mean civil time.

The mean synodic revolution of IV $\times$ 21 = 351ᵈ 19ʰ 47ᵐ 25ˢ.49492
$\times$ 22 = 368 13 52 32.42325

The terms of the arguments of the inequalities which increase uniformly with the time are based on the data of the following table :

Argument.	Terms.		Adopted value for first mean conj. 1880.		Motion for $i=21$.	Motion for $i=22$.
			R.	o.	o.	o.
1	$u_o - \pi_o$		11	8.4401	29.2317	30.6237
2	$U - \pi'$		11	29.2893	346.7584	3.2707
3	$U - u_o - 7°.43$		9	11.8840	317.5280	332.6484
4	$u_{III} - u_{IV} - 7°.40$		6	10.7138	33.7389	155.3456
5	$u_{II} - u_{IV} - 20°.50$		4	8.7027	356.8882	253.8829
6	$u_{IV} - \pi_{IV} - 5°.53$		2	12.3656	28.5616	29.9216
7	$u_{IV} - \pi_{III} - 5°.53$		1	19.9755	26.7248	27.9974
I	$u_{IV} - II - 6°.36$		0	29.3739	29.2311	30.6262
II	$u_{IV} - \Lambda_{IV} - 6°.36$		5	4.3479	29.9003	31.3241
III	$u_{IV} - \Lambda_{III} - 6°.36$		4	21.2573	31.6942	33.2035
IV	$u_{IV} - \Lambda_{II} - 6°.36$		0	10.8496	40.8682	42.8144

The remaining terms of the arguments depend on J, the great inequality of Jupiter ; and on ϕ, which is employed by DE DAMOISEAU to represent " les perturbations en longitude " of Jupiter.

Let there be

ϕ, the sum of the equations from BOUVARD's Tables* XIII—XXVI, } Longitude.
Σk, the sum of the constants added to these tables.

ϕ_1, the sum of the equations from BOUVARD's Tables XXVIII—XXXVI, } Radius vector.
Σk_1, the sum of the constants added to these tables.

I then understand from DE DAMOISEAU, Introduction, page (III), that

$$\phi = \phi' - \Sigma k,$$

that is, the "perturbations in longitude" (ϕ) do not include the great inequality of Jupiter.

It was found, however, that in order to form the complete arguments of the inequalities of the satellites, as DE DAMOISEAU appears to have done, it is necessary to add the great inequality. So that in the formulae for the arguments alone, given on pages (IV), (V), (VII), (IX) of the Introduction to the *Tables Écliptiques*, it is necessary to write $J + \phi$, instead of ϕ.

THE FORMATION OF TABLE III.

The Terms in ($J + \phi + \delta E$).—In continuing Table III of each satellite, the values of J, ϕ, ϕ_1, δr, δE, were computed from BOUVARD's Tables at half-year intervals, ϕ_1 being equal to $\phi'_1 - \Sigma k_1$. The results of this computation are presented in the following table. J and δE, columns the second and fifth, are expressed in centesimal arc. $J + \phi$, column the third, is expressed in sexagesimal arc.

$$\Sigma k = 22' 11''.5 \text{ (of centesimal arc),}$$
$$= 0°.19903 \text{ (of sexagesimal arc).}$$
$$\Sigma k_1 = 0.00730$$

Year and tenth.	J		$J + \phi$	$\phi_1 + \delta r$	δE	
			o.		o.	
1880.0 B	+ 29	56.6	+ 0.12934	− 0.00165	− 0	63.7
1880.5 B	29	49.2	0.14189	− 0.00318	− 0	19.1
1881.0	29	41.7	0.16411	− 0.00432	+ 0	28.3
1881.5	29	34.3	0.19297	− 0.00197	0	72.9
1882.0	+ 29	26.7	+ 0.22440	− 0.00499	+ 1	11.6

---

* Tables Astronomiques publiées par le Bureau des Longitudes de France, contenant les Tables de Jupiter, de Saturne et d'Uranus, construites d'après la Théorie de la Mécanique Céleste, par M. A. BOUVARD, Paris, 1821.

2

Year and tenth.	J	J + φ	φ₁ + δr	δ E
1882.0	+ 29° 26″.7	+ 0.22110	− 0.00199	+ 1′ 11″.6
1882.5	29 19.2	0.25309	− 0.00451	1 40.3
1883.0	29 11.6	0.27666	− 0.00363	1 57.9
1883.5	29 3.9	0.29263	− 0.00248	1 63.5
1884.0 B	28 96.3	0.30115	− 0.00120	1 57.5
1884.5 B	28 88.6	0.30277	+ 0.00003	1 41.5
1885.0	+ 28 80.9	+ 0.29894	+ 0.00115	+ 1 16.0
1885.5	28 73.2	0.29088	0.00207	0 85.7
1886.0	28 65.3	0.27967	0.00275	0 49.4
1886.5	28 57.5	0.26640	0.00322	+ 0 10.8
1887.0	28 49.6	0.25174	0.00343	− 0 29.3
1887.5	+ 28 41.8	+ 0.23672	+ 0.00331	− 0 67.8
1888.0 B	28 33.9	0.22196	0.00295	− 1 4.0
1888.5 B	28 25.9	0.20897	0.00230	− 1 31.8
1889.0	28 17.9	0.19816	0.00142	− 1 58.6
1889.5	28 9.8	0.19210	+ 0.00044	− 1 72.6
1890.0	+ 28 1.8	+ 0.19063	− 0.00068	− 1 75.5
1890.5	27 93.7	0.19467	− 0.00173	− 1 65.6
1891.0	27 85.6	0.20136	− 0.00270	− 1 42.0
1891.5	27 77.4	0.21906	− 0.00337	− 1 5.9
1892.0 B	27 69.2	0.23743	− 0.00377	− 0 59.5
1892.5 B	+ 27 61.0	+ 0.25788	− 0.00378	− 0 6.9
1893.0	27 52.7	0.27843	− 0.00348	+ 0 47.1
1893.5	27 44.4	0.29665	− 0.00292	0 96.9
1894.0	27 36.1	0.31058	− 0.00210	1 36.1
1894.5	27 27.7	0.31931	− 0.00122	1 67.5
1895.0	+ 27 19.3	+ 0.32211	− 0.00026	+ 1 83.5
1895.5	27 10.9	0.31916	+ 0.00063	1 86.0
1896.0 B	27 2.4	0.31155	0.00143	1 75.8
1896.5 B	26 93.9	0.30063	0.00203	1 54.4
1897.0	26 85.1	0.28779	0.00256	1 23.9
1897.5	+ 26 76.8	+ 0.27445	+ 0.00285	+ 0 86.5
1898.0	26 68.2	0.26148	0.00296	+ 0 43.7
1898.5	26 59.6	0.24971	0.00282	− 0 0.4
1899.0	26 50.9	0.23996	0.00247	− 0 45.6
1899.5	26 42.2	0.23291	0.00185	− 0 89.0
1900.0	+ 26 33.5	+ 0.22939	+ 0.00107	− 1 28.3

The mean synodic revolutions of the satellites are

$$I = 1.769860478875$$
$$II = 3.554094157794$$
$$III = 7.166387201355$$
$$IV = 16.753552411222,$$

which are fifteen times the factors for ($J + φ + δE$). The data already presented suffice for the computation of Table III of the first satellite.

The Terms in ($5 ñ − 2 u_o$).—These terms of Table III, satellites II, III, IV, are functions of the longitudes of Jupiter and Saturn.

$$1880.0 \qquad (5 ñ − 2 u_o − 34°.512) = 84°.068$$
$$\text{Daily motion of the angle } (5 ñ − 2 u_o) = 0°.001039596$$

During the period 1880—1900 this angle plus the constant is so near 90° that its sine varies very slowly, and it will be sufficient to compute the terms involving its sine for every fifth year.

	II.	III.	III.
	$+(^*)0^s.952 \sin(5\tilde{n} - 2v_0 - 34^\circ.542)$	$+(^*)2^s.823 \sin(5\tilde{n} - 2v_0 - 34^\circ.542)$	$+(^*)15^s.581 \sin(5\tilde{n} - 2v_0 - 34^\circ.542)$
1880.0	$+0.95$	$+2.81$	$+15.50$
1885.0	$+0.95$	$+2.82$	$+15.54$
1890.0	$+0.95$	$+2.82$	$+15.57$
1895.0	$+0.95$	$+2.82$	$+15.58$
1900.0	$+0.95$	$+2.82$	$+15.57$

The Terms in ($\text{II} - \Lambda_\text{II}$), ($\text{II} - \Lambda_\text{III}$), *etc.*— These terms of Table III, satellites II, III, IV, are functions of the longitude of the ascending node of the equator of Jupiter on its orbit, and of the longitudes of the ascending nodes of the orbits of these satellites on their fixed planes.

$$1880.0 \quad (\text{II} - \Lambda_\text{II}) = 341^\circ.499$$
$$1880.0 \quad (\text{II} - \Lambda_\text{III}) = 111.875$$
$$1880.0 \quad (\text{II} - \Lambda_\text{IV}) = 124.964$$

$$\text{Daily motion of the angle } (\text{II} - \Lambda_\text{II}) = 0^\circ.03306928964$$
$$\text{Daily motion of the angle } (\text{II} - \Lambda_\text{III}) = 0.00699245908$$
$$\text{Daily motion of the angle } (\text{II} - \Lambda_\text{IV}) = 0.00189341824$$

The angle ($\text{II} - \Lambda_\text{IV}$) changes so slowly that the computation of the term involving its sine for every fifth year will suffice. The term in ($\text{II} - \Lambda_\text{II}$) has been computed partly at intervals of one, and partly at intervals of two years. The term depending on ($\text{II} - \Lambda_\text{II}$) has been computed at half-year intervals.

Year and tenth.	II.		III.		IV.	
	$-(^*)9^s.731 \sin(\text{II} - \Lambda_\text{II})$		$-(^*)5^s.775 \sin(\text{II} - \Lambda_\text{III})$		$+16^s.694 \sin(\text{II} - \Lambda_\text{IV})$	
	″	Diff.	″	Diff.	″	Diff.
1880.0 B	$+\ 3.09$	99	$-\ 5.36$		$+\ 13.68$	
1880.5 B	2.10	102		10		
1881.0	1.08	102	$-\ 5.26$			
1881.5	$+\ 0.06$	102		11		
1882.0	$-\ 0.96$	101	$-\ 5.15$			
1882.5	$-\ 1.97$	100		13		60
1883.0	$-\ 2.97$	95	$-\ 5.02$			
1883.5	$-\ 3.92$	92		13		
1884.0 B	$-\ 4.81$	86	$-\ 4.89$			
1884.5 B	$-\ 5.70$	80				
1885.0	$-\ 6.50$	73		29	$+\ 13.08$	
1885.5	$-\ 7.23$	64				
1886.0	$-\ 7.87$	56	$-\ 4.60$			
1886.5	$-\ 8.43$	46				
1887.0	$-\ 8.89$	37		33		
1887.5	$-\ 9.26$	26				65
1888.0 B	$-\ 9.52$	16	$-\ 4.27$			
1888.5 B	$-\ 9.68$	5				
1889.0	$-\ 9.73$	6		37		
1889.5	$-\ 9.67$	16				
1890.0	$-\ 9.51$	27	$-\ 3.90$		$+\ 12.43$	
1890.5	$-\ 9.21$	38				
1891.0	$-\ 8.86$	47		39		
1891.5	$-\ 8.39$	56				
1892.0 B	$-\ 7.83$	65	$-\ 3.51$			70

(*) This inequality is not given by the theory of Laplace.

Year and tenth.	II. $-(*)9^s.731 \sin(\Pi - A_{II})$		III. $-(*)5^s.775 \sin(\Pi - A_{III})$		IV. $+16^s.694 \sin(\Pi - A_{IV})$	
	s	Diff.	s	Diff.	s	Diff.
1892.0 B	− 7.83	65	− 3.51			
1892.5 B	− 7.18	74				70
1893.0	− 6.44	80		42		
1893.5	− 5.64	87				
1894.0	− 4.77	91	− 3.09			
1894.5	− 3.86	96				
1895.0	− 2.90	100		44	+ 11.73	
1895 5	− 1.90	101				
1896.0 B	− 0.89	103	− 2.65			
1896.5 B	+ 0.14	102				
1897.0	1.16	101		47		
1897.5	+ 2.17	99				73
1898.0	3.16	95	− 2.18			
1898.5	4.11	90				
1899.0	5.01	85		49		
1899.5	5.86	79				
1900.0	+ 6.65		− 1.69		+ 11 00	

My values of the "perturbations of Jupiter and other inequalities," Table III, for the epoch 1860.0, do not agree precisely with those given by De Damoiseau.

For farther comparison with his tables, I have computed, in this way, Table III of each satellite complete for the years 1878 and 1879; and while the method is probably the one employed by De Damoiseau, it does not suffice to reproduce exactly his values of the perturbations (Table III) for these years.

The corrections necessary to reduce his values to such as have been computed in the manner indicated are as follow:

Satellite	I.	II.	III.	IV.
1878.0	+ 1.7	+ 2.5	+ 7.2	+ 16.2
1879.0	2.1	3.1	9.2	20.7
1880.0	+ 2.1	+ 3.5	+ 8.6	+ 19.5

The discrepancy is traceable to the term

$$(J + \phi + \delta E),$$

and δE appears to have been reduced by De Damoiseau from the epoch 1714.0; while the epoch of δE of Bouvard's Tables is 1800.0.

The differences alluded to are less than the accidental errors of observation, and may be disregarded.

For precepts for the use of the tables of the eclipses, the computer is referred to the Introduction to the *Tables Écliptiques*, pages (x)—(xvii).

Table A has been adapted from advance sheets of the *American Ephemeris and Nautical Almanac* for 1880, and gives the longitudes of Observatories, Paris being the prime meridian. West longitudes are considered positive. To reduce the tabular instant of an eclipse to the mean solar astronomic time of any meridian having a longitude λ from Paris, it is necessary to apply the correction

$$- (\lambda + 12^h).$$

The Tables of the Configurations.—Of the Tables for computing the configurations, Table I of each satellite alone requires extension.

(*) This inequality is not given by the theory of Laplace.

The data for the continuation of the column "Mean longitude," are given by DE DAMOISEAU, Introduction, page (III). Whence there is derived the

$$
\begin{array}{ll}
\text{Motion of } u_{\text{I}} \text{ in 365 days,} & 113\overset{\circ}{.}48258 \\
\text{Motion of } u_{\text{I}} \text{ in 366 days,} & 316.97157 \\
\text{Motion of } u_{\text{II}} \text{ in 365 days,} & 281.78815 \\
\text{Motion of } u_{\text{II}} \text{ in 366 days,} & 23.16291 \\
\text{Motion of } u_{\text{III}} \text{ in 365 days,} & 5.94095 \\
\text{Motion of } u_{\text{III}} \text{ in 366 days,} & 56.25860 \\
\text{Motion of } u_{\text{IV}} \text{ in 365 days,} & 313.45494 \\
\text{Motion of } u_{\text{IV}} \text{ in 366 days,} & 335.02605.
\end{array}
$$

There was adopted, for the epoch 1880, January 1, Paris mean midnight,

$$
\begin{array}{ll}
u_{\text{I}} = & 6^{\text{s}} \ 9\overset{\circ}{.}80 \\
u_{\text{II}} = & 4 \ 3.87 \\
u_{\text{III}} = & 6 \ 0.92 \\
u_{\text{IV}} = & 11 \ 17.40
\end{array}
$$

DE DAMOISEAU gives no explanation of the method of formation of the arguments of these tables. I have, therefore, continued them by induction.

For precepts for the use of the tables of the configurations, the computer is referred to the *Tables Écliptiques*, pages (193)—(199).

ERRORS IN DE DAMOISEAU'S *Tables Écliptiques, etc.*—Through the courtesy of Mr. J. R. HIND, F. R. S., the Superintendent of the *British Nautical Almanac*, and of Professor E. O. KENDALL, of the University of Pennsylvania, I have been enabled to make the appended list of errors and corrections more complete than it would otherwise have been.

TABLES.

TABLE I.   Epochs of the Mean Conjunctions

Years.	Mean Conjunctions. Days and parts of a day, Paris mean civil time.	Fraction of year.	1	2	3	4
1880 B	Jan. 1 18 8 15.0	0.002	11 8.750	11 29.91	9 12.32	4 24.05
1881	2 2 48 15.7	0.003	0 9.197	0 0.99	8 12.04	3 24.76
1882	1 16 59 40.4	0.002	1 9.488	0 0.34	7 11.93	8 24.74
1883	1 7 11 5.2	0.001	2 9.779	11 29.68	6 10.92	1 24.72
1884 B	2 15 51 5.8	0.004	3 10.217	0 0.76	5 11.55	0 25.42
1885	1 6 2 30.6	0.001	4 10.508	0 0.10	4 10.60	5 25.38
1886	2 14 42 31.3	0.004	5 10.946	0 1.19	3 11.27	4 26.05
1887	2 4 53 56.0	0.003	6 11.237	0 0.53	2 10.34	9 25.99
1888 B	1 19 5 20.8	0.002	7 11.528	11 29.87	1 9.42	2 25.93
1889	2 3 45 21.5	0.003	8 11.966	0 0.96	0 10.10	1 26.61
1890	1 17 56 46.2	0.002	9 12.257	0 0.30	11 9.15	6 26.56
1891	1 8 8 11.0	0.001	10 12.548	11 29.64	10 8.19	11 26.52
1892 B	2 16 48 11.7	0.005	11 12.986	0 0.72	9 8.80	10 27.22
1893	1 6 59 36.4	0.001	0 13.277	0 0.07	8 7.81	3 27.20
1894	2 15 39 37.1	0.004	1 13.715	0 1.15	7 8.43	2 27.00
1895	2 5 51 1.8	0.003	2 14.006	0 0.49	6 7.47	7 27.86
1896 B	1 20 2 26.6	0.002	3 14.297	11 29.83	5 6.53	0 27.81
1897	2 4 42 27.3	0.003	4 14.735	0 0.92	4 7.20	11 28.48
1898	1 18 53 52.0	0.002	5 15.026	0 0.26	3 6.28	4 28.43
1899	1 9 5 16.8	0.001	6 15.316	11 29.60	2 5.35	9 28.37
1900	2 17 45 17.5	0.005	7 15.754	0 0.69	1 6.00	8 29.05

and the Arguments of the Inequalities.

YEARS.	5	6	7	8	9	I	II	III
1880 B	4 6.1	1 20.3	2 12.6	7 28.0	7 5.6	1 0.35	0 11.9	4 22.2
1881	2 22.2	2 18.2	3 12.3	5 1.6	4 7.2	2 0.82	1 24.5	5 25.3
1882	4 7.2	3 16.0	4 12.0	2 3.8	1 7.5	3 1.18	3 6.9	6 28.2
1883	5 22.1	4 13.7	5 11.6	11 6.0	10 7.8	4 1.52	4 19.3	8 1.1
1884 B	4 8.2	5 11.6	6 11.1	8 9.6	7 9.5	5 1.99	6 1.9	9 4.1
1885	5 23.1	6 9.2	7 11.0	5 11.8	4 9.8	6 2.28	7 14.2	10 6.9
1886	4 9.1	7 7.1	8 10.7	2 15.3	1 11.1	7 2.70	8 26.8	11 9.9
1887	5 24.0	8 4.7	9 10.3	11 17.5	10 11.7	8 2.97	10 9.1	0 12.7
1888 B	7 8.9	9 2.4	10 9.9	8 19.7	7 12.0	9 3.21	11 21.4	1 15.6
1889	5 25.0	10 0.2	11 9.6	5 23.3	4 13.7	10 3.65	1 4.0	2 18.5
1890	7 9.9	10 27.9	0 9.2	2 25.5	1 14.0	11 3.91	2 16.3	3 21.4
1891	8 24.8	11 25.6	1 8.8	11 27.7	10 14.3	0 4.25	3 28.7	4 24.2
1892 B	7 10.9	0 23.5	2 8.6	9 1.3	7 15.9	1 4.72	5 11.2	5 27.3
1893	8 25.8	1 21.2	3 8.2	6 3.5	4 16.2	2 5.06	6 23.6	7 0.1
1894	7 11.9	2 19.1	4 8.0	3 7.0	1 17.8	3 5.53	8 6.2	8 3.2
1895	8 26.8	3 16.8	5 7.6	0 9.2	10 18.2	4 5.84	9 18.6	9 6.0
1896 B	10 11.8	4 11.5	6 7.2	9 11.4	7 18.5	5 6.12	11 0.9	10 8.9
1897	8 27.8	5 12.3	7 6.9	6 15.0	4 20.1	6 6.54	0 13.5	11 11.8
1898	10 12.7	6 9.9	8 6.5	3 17.2	1 20.1	7 6.81	1 25.8	0 14.7
1899	11 27.6	7 7.6	9 6.0	0 19.4	10 20.7	8 7.08	3 8.1	1 17.5
1900	10 13.6	8 5.4	10 5.8	9 23.0	7 22.4	9 7.51	4 20.7	2 20.5

THE FIRST SATELLITE.

TABLE III.   Perturbations of Jupiter and other Inequalities,

$$0.1179907 \,(J + \phi + \delta E) + 493.2 \,(\delta_1 + \delta r).$$

Years and tenths.	Perturb.	Diff.	Years and tenths.	Perturb.	Diff.	Years and tenths.	Perturb.	Diff.	Years and tenths.	Perturb.	Diff.
	m s	s		m s	s		m s	s		m s	s
1880.0	0 51.7	+0.8	1883.0	2 1.7	+1.9	1886.0	2 2.1	-1.3	1889.0	1 19.0	-1.0
1	0 52.5	1.1	1	2 3.6	1.6	1	2 0.8	-1.4	1	1 18.0	-0.8
2	0 53.6	1.3	2	2 5.2	1.6	2	1 59.4	-1.4	2	1 17.2	-0.8
3	0 54.9	1.4	3	2 6.8	1.4	3	1 58.0	-1.4	3	1 16.4	-0.6
4	0 56.3	+1.7	4	2 8.2	+1.2	4	1 56.6	-1.5	4	1 15.8	-0.6
5	0 58.0	1.8	5	2 9.4	1.1	5	1 55.1	-1.5	5	1 15.2	-0.5
6	0 59.8	2.0	6	2 10.5	0.9	6	1 53.6	-1.5	6	1 14.7	-0.3
7	1 1.8	2.1	7	2 11.4	0.8	7	1 52.1	-1.5	7	1 14.4	-0.3
8	1 3.9	2.3	8	2 12.2	0.6	8	1 50.6	-1.5	8	1 14.1	-0.2
9	1 6.2	+2.5	9	2 12.8	+0.5	9	1 49.1	-1.6	9	1 13.9	0.0
1881.0	1 8.7	2.5	1884.0	2 13.3	0.4	1887.0	1 47.5	-1.6	1890.0	1 13.9	+0.1
1	1 11.2	2.7	1	2 13.7	0.3	1	1 45.9	-1.6	1	1 14.0	0.2
2	1 13.9	2.7	2	2 14.0	0.1	2	1 44.3	-1.6	2	1 14.2	0.3
3	1 16.6	2.9	3	2 14.1	0.0	3	1 42.7	-1.6	3	1 14.5	0.5
4	1 19.5	+2.9	4	2 14.1	-0.1	4	1 41.1	-1.6	4	1 15.0	+0.5
5	1 22.4	3.0	5	2 14.0	-0.2	5	1 39.5	-1.6	5	1 15.5	0.7
6	1 25.4	2.9	6	2 13.8	-0.3	6	1 37.9	-1.5	6	1 16.2	0.8
7	1 28.3	3.1	7	2 13.5	-0.4	7	1 36.4	-1.6	7	1 17.0	0.9
8	1 31.4	2.8	8	2 13.1	-0.5	8	1 34.8	-1.5	8	1 17.9	1.0
9	1 34.2	+2.9	9	2 12.6	-0.6	9	1 33.3	-1.5	9	1 18.9	+1.1
1882.0	1 37.1	2.8	1885.0	2 12.0	-0.7	1888.0	1 31.8	-1.5	1891.0	1 20.0	1.3
1	1 39.9	2.9	1	2 11.3	-0.8	1	1 30.3	-1.5	1	1 21.3	1.4
2	1 42.8	2.7	2	2 10.5	-0.8	2	1 28.8	-1.4	2	1 22.7	1.5
3	1 45.5	2.7	3	2 9.7	-0.9	3	1 27.4	-1.4	3	1 24.2	1.5
4	1 48.2	+2.5	4	2 8.8	-1.0	4	1 26.0	-1.3	4	1 25.7	+1.6
5	1 50.7	2.5	5	2 7.8	-1.0	5	1 24.7	-1.3	5	1 27.3	1.8
6	1 53.2	2.3	6	2 6.8	-1.1	6	1 23.4	-1.2	6	1 29.1	1.8
7	1 55.5	2.2	7	2 5.7	-1.2	7	1 22.2	-1.1	7	1 30.9	1.9
8	1 57.7	2.1	8	2 4.5	-1.2	8	1 21.1	-1.1	8	1 32.8	1.9
9	1 59.8	+1.9	9	2 3.3	-1.2	9	1 20.0	-1.0	9	1 34.7	+2.0
1883.0	2 1.7		1886.0	2 2.1		1889.0	1 19.0		1892.0	1 36.7	

TABLE III. Perturbations of Jupiter and other Inequalities.

$$0.1170907\,(J + \phi + \delta E) + 493.2\,(\phi_1 + \delta r).$$

Years and tenths.	Perturb.	Diff.	Years and tenths.	Perturb.	Diff.	Years and tenths.	Perturb.	Diff.	Years and tenths.	Perturb.	Diff.
	m s	s		m s	s		m s	s		m s	s
1892.0	1 36.7	+2.1	1894.0	2 16.1	+1.3	1896.0	2 19.8	-0.9	1898.0	1 54.2	-1.4
1	1 38.8	2.1	1	2 17.4	1.2	1	2 18.9	-1.0	1	1 52.8	-1.4
2	1 40.9	2.2	2	2 18.6	1.1	2	2 17.9	-1.0	2	1 51.4	-1.3
3	1 43.1	2.1	3	2 19.7	0.9	3	2 16.9	-1.1	3	1 50.1	-1.4
4	1 45.2	+2.2	4	2 20.6	+0.8	4	2 15.8	-1.2	4	1 48.7	-1.3
5	1 47.4	2.2	5	2 21.4	0.7	5	2 14.6	-1.2	5	1 47.4	-1.2
6	1 49.6	2.2	6	2 22.1	0.6	6	2 13.4	-1.2	6	1 46.2	-1.3
7	1 51.8	2.2	7	2 22.7	0.5	7	2 12.2	-1.3	7	1 44.9	-1.2
8	1 54.0	2.2	8	2 23.2	0.3	8	2 10.9	-1.3	8	1 43.7	-1.2
9	1 56.2	+2.1	9	2 23.5	+0.2	9	2 9.6	-1.3	9	1 42.5	-1.1
1893.0	1 58.3	2.1	1895.0	2 23.7	+0.1	1897.0	2 8.3	-1.4	1899.0	1 41.4	-1.1
1	2 0.4	2.1	1	2 23.8	0.0	1	2 6.9	-1.4	1	1 40.3	-1.0
2	2 2.5	2.0	2	2 23.8	-0.2	2	2 5.5	-1.4	2	1 39.3	-1.0
3	2 4.5	1.9	3	2 23.6	-0.3	3	2 4.1	-1.4	3	1 38.3	-1.0
4	2 6.4	+1.8	4	2 23.3	-0.3	4	2 2.7	-1.4	4	1 37.3	-0.9
5	2 8.2	1.8	5	2 23.0	-0.5	5	2 1.3	-1.4	5	1 36.4	-0.8
6	2 10.0	1.7	6	2 22.5	-0.5	6	1 59.9	-1.4	6	1 35.6	-0.7
7	2 11.7	1.6	7	2 22.0	-0.7	7	1 58.5	-1.5	7	1 34.9	-0.7
8	2 13.3	1.5	8	2 21.3	-0.7	8	1 57.0	-1.4	8	1 34.2	-0.6
9	2 14.8	+1.3	9	2 20.6	-0.8	9	1 55.6	-1.4	9	1 33.6	-0.5
1894.0	2 16.1		1896.0	2 19.8		1898.0	1 54.2		1900.0	1 33.1	

TABLE I.   Epochs of the Mean Conjunctions

| YEARS. | MEAN CONJUNCTIONS. | | 1 | 2 | 3 | 4 |
	Days and parts of a day, Paris mean civil time.	Fraction of year.				
1880 B	Jan. 3 4 1 1.0	0.006	11 8.879	0 1.34	9 13.63	1 24.63
1881	3 5 44 15.7	0.006	0 9.293	0 2.14	8 13.98	0 25.21
1882	4 7 27 30.5	0.009	1 9.707	0 2.94	7 14.31	11 25.81
1883	1 19 52 51.5	0.002	2 9.826	0 0.23	6 11.43	4 24.94
1884 B	2 21 36 6.2	0.005	3 10.240	0 1.03	5 11.80	3 25.52
1885	2 23 19 20.9	0.005	4 10.654	0 1.83	4 12.18	2 26.08
1886	4 1 2 35.6	0.008	5 11.068	0 2.64	• 3 12.59	1 26.64
1887	1 13 27 56.6	0.001	6 11.187	11 29.93	· 2 9.80	6 25.72
1888 B	2 15 11 11.4	0.004	7 11.601	0 0.73	1 10.21	5 26.27
1889	2 16 54 26.1	0.005	8 12.015	0 1.53	0 10.62	4 26.83
1890	3 18 37 40.8	0.008	9 12.429	0 2.33	11 11.02	3 27.39
1891	1 7 3 1.8	0.001	10 12.548	11 29.63	10 8.18	8 26.50
1892 B	2 8 46 16.6	0.004	11 12.961	0 0.43	9 8.53	7 27.08
1893	2 10 29 31.3	0.004	0 13.375	0 1.23	8 8.88	6 27.67
1894	3 12 12 46.0	0.007	1 13.789	0 2.03	7 9.23	5 28.25
1895	4 13 56 0.7	0.010	2 14.203	0 2.83	6 9.61	4 28.82
1896 B	2 2 21 21.7	0.003	3 14.322	0 0.13	5 6.80	9 27.92
1897	2 4 4 36.5	0.003	4 14.736	0 0.93	4 7.21	8 28.47
1898	3 5 47 51.3	0.006	5 15.150	0 1.73	3 7.62	7 29.02
1899	4 7 31 5.9	0.009	6 15.563	0 2.53	2 8.03	6 29.58
1900	1 19 56 26.9	0.002	7 15.682	11 29.83	1 5.22	11 28.67

and the Arguments of the Inequalities.

YEARS.	5	6	7	8	I	II	III	IV
1880 B	1 20.2	2 12.7	7 29.1	7 6.6	1 0.46	0 12.05	4 22.1	5 5.4
1881	2 18.1	3 12.4	5 2.4	4 8.1	2 0.92	1 24.61	5 25.4	6 6.6
1882	3 15.9	4 12.2	2 5.8	1 9.5	3 1.39	3 7.19	6 28.4	7 7.8
1883	4 13.5	5 11.7	11 6.4	10 8.2	4 1.57	4 19.35	8 1.1	8 8.6
1884 B	5 11.3	6 11.4	8 9.8	7 9.7	5 2.01	6 1.90	9 4.1	9 9.7
1885	6 9.2	7 11.2	5 13.1	4 11.1	6 2.42	7 14.42	10 7.1	10 10.8
1886	7 6.9	8 10.8	2 16.4	1 12.5	7 2.82	8 26.93	11 10.0	11 11.9
1887	8 4.4	9 10.2	11 17.1	10 11.3	8 2.92	10 9.01	0 12.7	0 12.7
1888 B	9 2.2	10 9.9	8 20.4	7 12.7	9 3.30	11 21.50	1 15.6	1 13.8
1889	10 0.0	11 9.6	5 23.7	4 11.1	10 3.70	1 4.00	2 18.6	2 11.9
1890	10 27.8	0 9.3	2 27.0	1 15.5	11 4.11	2 16.52	3 21.6	3 16.0
1891	11 25.4	1 8.8	11 27.7	10 14.3	0 4.24	3 28.64	4 24.2	4 16.8
1892 B	0 23.2	2 8.5	9 1.0	7 15.7	1 4.69	5 11.20	5 27.2	5 18.0
1893	1 21.0	3 8.3	6 4.4	4 17.1	2 5.15	6 23.76	7 0.2	6 19.1
1894	2 18.9	4 8.0	3 7.7	1 18.5	3 5.60	8 6.32	8 3.3	7 20.2
1895	3 16.7	5 7.8	0 11.0	10 19.9	4 6.03	9 18.85	9 6.2	8 21.4
1896 B	4 14.2	6 7.2	9 11.7	7 18.7	5 6.14	11 0.95	10 8.9	9 22.2
1897	5 12.0	7 6.9	6 15.0	4 20.1	6 6.54	0 13.45	11 11.8	10 23.3
1898	6 9.8	8 6.6	3 18.3	1 21.5	7 6.93	1 25.95	0 14.8	11 24.3
1899	7 7.6	9 6.3	0 21.6	10 22.9	8 7.32	3 8.45	1 17.8	0 25.4
1900	8 5.1	10 5.7	9 22.3	7 21.7	9 7.43	4 20.55	2 20.4	1 26.2

THE SECOND SATELLITE.

TABLE III.   Perturbations of Jupiter and other Inequalities.

$$0.2369396\,(J + \phi + \delta E) + 493.2\,(\phi_1 + \delta r) + (^*)\,0.952\,\sin(5\bar{u} - 2u_0 - 34.542) - (^*)\,9.731\,\sin(H - \Lambda_{II}).$$

Years and tenths.	Perturb.	Diff.	Years and tenths.	Perturb.	Diff.	Years and tenths.	Perturb.	Diff.	Years and tenths.	Perturb.	Diff.
	m s	s		m s	s		m s	s		m s	s
1880.0	1 48.7	+1.7	1883.0	4 4.3	+3.3	1886.0	3 56.8	−2.8	1889.0	2 29.1	−1.8
1	1 50.4	2.0	1	4 7.6	3.1	1	3 54.0	−2.9	1	2 27.3	−1.6
2	1 52.4	2.5	2	4 10.7	2.7	2	3 51.1	−2.9	2	2 25.7	−1.4
3	1 54.9	2.9	3	4 13.4	2.5	3	3 48.2	−3.0	3	2 24.3	−1.2
4	1 57.8	+3.3	4	4 15.9	+2.2	4	3 45.2	−3.1	4	2 23.1	−1.0
5	2 1.1	3.6	5	4 18.1	1.7	5	3 42.1	−3.1	5	2 22.1	−0.8
6	2 4.7	3.9	6	4 19.8	1.6	6	3 39.0	−3.1	6	2 21.3	−0.6
7	2 8.6	4.2	7	4 21.4	1.3	7	3 35.9	−3.2	7	2 20.7	−0.4
8	2 12.8	4.5	8	4 22.7	1.0	8	3 32.7	−3.2	8	2 20.3	−0.2
9	2 17.3	+4.8	9	4 23.7	+0.8	9	3 29.5	−3.3	9	2 20.1	+0.1
1881.0	2 22.1	5.0	1884.0	4 24.5	0.4	1887.0	3 26.2	−3.2	1890.0	2 20.2	0.3
1	2 27.1	5.2	1	4 24.9	+0.2	1	3 23.0	−3.3	1	2 20.5	0.6
2	2 32.3	5.4	2	4 25.1	0.0	2	3 19.7	−3.3	2	2 21.1	0.8
3	2 37.7	5.6	3	4 25.1	−0.2	3	3 16.4	−3.3	3	2 21.9	1.0
4	2 43.3	+5.7	4	4 24.9	−0.5	4	3 13.1	−3.2	4	2 22.9	+1.3
5	2 49.0	5.7	5	4 24.4	−0.7	5	3 9.9	−3.2	5	2 24.2	1.6
6	2 54.7	5.7	6	4 23.7	−0.9	6	3 6.7	−3.2	6	2 25.8	1.7
7	3 0.4	5.7	7	4 22.8	−1.1	7	3 3.5	−3.1	7	2 27.5	2.0
8	3 6.1	5.7	8	4 21.7	−1.3	8	3 0.4	−3.1	8	2 29.5	2.2
9	3 11.8	+5.6	9	4 20.4	−1.5	9	2 57.3	−3.1	9	2 31.7	+2.5
1882.0	3 17.4	5.5	1885.0	4 18.9	−1.6	1888.0	2 54.2	−2.9	1891.0	2 34.2	2.7
1	3 22.9	5.5	1	4 17.3	−1.8	1	2 51.3	−2.9	1	2 36.9	2.9
2	3 28.4	5.9	2	4 15.5	−1.9	2	2 48.4	−2.8	2	2 39.8	3.1
3	3 34.6	5.1	3	4 13.6	−2.0	3	2 45.6	−2.7	3	2 42.9	3.3
4	3 39.7	+4.8	4	4 11.6	−2.2	4	2 42.9	−2.7	4	2 46.2	+3.4
5	3 43.5	4.7	5	4 9.4	−2.3	5	2 40.2	−2.5	5	2 49.6	3.7
6	3 48.2	4.4	6	4 7.1	−2.4	6	2 37.7	−2.3	6	2 53.3	3.8
7	3 52.6	4.1	7	4 4.7	−2.6	7	2 35.4	−2.3	7	2 57.1	3.9
8	3 56.7	3.9	8	4 2.1	−2.6	8	2 33.1	−2.1	8	3 1.0	4.0
9	4 0.6	+3.7	9	3 59.5	−2.7	9	2 31.0	−1.9	9	3 5.0	+4.2
1883.0	4 4.3		1886.0	3 56.8		1889.0	2 29.1		1892.0	3 9.2	

TABLE III.   Perturbations of Jupiter and other Inequalities.

$$0.2369396 (J + \phi + \delta E) + 493.2 (\phi_1 + \delta r) + (^*) 0.952 \sin (5\bar{a} - 2v_0 - 34.542) - (^*) 9.731 \sin (\text{II} - A_{\text{II}}).$$

Years and tenths.	Perturb.	Diff.	Years and tenths.	Perturb.	Diff.	Years and tenths.	Perturb.	Diff.	Years and tenths.	Perturb.	Diff.
	m s	s		m s	s		m s	s		m s	s
1892.0	3 9.2	+4.3	1894.0	4 30.6	+2.7	1896.0	4 40.0	-1.7	1898.0	3 52.0	-2.6
1	3 13.5	4.4	1	4 34.3	2.4	1	4 38.3	-1.8	1	3 49.4	-2.6
2	3 17.9	4.4	2	4 35.7	2.2	2	4 36.5	-1.9	2	3 46.8	-2.5
3	3 22.3	4.5	3	4 37.9	2.0	3	4 34.6	-2.1	3	3 44.3	-2.5
4	3 26.8		4	4 39.9		4	4 32.5		4	3 41.8	
		+4.5			+1.8			-2.1			-2.4
5	3 31.3	4.5	5	4.41.7	1.5	5	4 30.4	-0.3	5	3 39.4	-2.3
6	3 35.8	4.8	6	4 43.2	1.3	6	4 28.1	-2.3	6	3 37.1	-2.3
7	3 40.4	4.5	7	4 44.5	1.0	7	4 25.8	-2.4	7	3 34.8	-2.3
8	3 44.9	4.5	8	4 45.5	0.7	8	4 23.1	-0.5	8	3 32.5	-2.1
9	3 49.4		9	4 46.2		9	4 20.9		9	3 30.4	
		+4.1			+0.5			-2.5			-2.0
1893.0	3 53.8	4.4	1895.0	4 46.7	+0.3	1897.0	4 18.4	-2.6	1899.0	3 28.4	-2.0
1	3 58.2	4.2	1	4 47.0	+0.1	1	4 15.8	-2.6	1	3 26.4	-1.9
2	4 2.4	4.1	2	4 47.1	-0.2	2	4 13.2	-2.6	2	3 24.5	-1.8
3	4 6.5	4.0	3	4 46.9	-0.4	3	4 10.6	-2.6	3	3 22.7	-1.6
4	4 10.5		4	4 46.5		4	4 8.0		4	3 21.1	
		+3.8			-0.6			-2.7			-1.5
5	4 14.3	3.6	5	4 45.9	-0.9	5	4 5.3	-2.7	5	3 19.6	-1.5
6	4 17.9	3.5	6	4 45.0	-1.0	6	4 2.6	-2.7	6	3 18.1	-1.3
7	4 21.4	3.3	7	4 44.0	-1.2	7	3 59.9	-0.6	7	3·16.8	-1.1
8	4 24.7	3.6	8	4 42.8	-1.3	8	3 57.3	-2.7	8	3 15.7	-0.9
9	4 27.7		9	4 41.5		9	3 54.6		9	3 14.8	
		+2.9			-1.5			-2.6			-0.9
1894.0	4 30.6		1896.0	4 40.0		1898.0	3 52.0		1900.0	3 13.9	

TABLE I.   Epochs of the Mean Conjunctions

YEARS.	MEAN CONJUNCTIONS.		1	2	3	4	5
	Days and parts of a day, Paris mean civil time.	Fraction of year.					
	Jan. h m s		s. °	s. °	s. °	s. °	s. °
1880 B	4 6 13 11.0	0.009	11 8.974	0 2.46	9 14.66	3 20.2	9 17.6
1881	3 17 52 39.6	0.008	0 9.339	0 2.68	8 14.48	1 20.8	11 24.1
1882	4 5 32 8.2	0.009	1 9.705	0 2.91	7 14.28	11 21.6	2 0.6
1883	4 17 11 36.7	0.010	2 10.070	0 3.13	6 11.09	9 22.2	4 7.1
1884 B	5 4 51 5.3	0.012	3 10.435	0 3.35	5 13.92	7 22.9	6 13.5
1885	4 16 30 33.9	0.010	4 10.801	0 3.58	4 13.78	5 23.6	8 20.0
1886	5 4 10 2.4	0.011	5 11.166	0 3.80	3 13.66	3 24.2	10 26.4
1887	5 15 49 31.0	0.013	6 11.531	0 4.02	2 13.54	1 24.8	1 2.8
1888 B	6 3 28 59.6	0.014	7 11.897	0 4.24	1 13.43	11 25.4	3 9.3
1889	5 15 8 28.1	0.013	8 12.262	0 4.47	0 13.31	9 26.1	5 15.7
1890	6 2 47 56.7	0.014	9 12.627	0 4.69	11 13.17	7 26.7	7 22.2
1891	6 14 27 25.3	0.015	10 12.993	0 4.91	10 13.02	5 27.4	9 28.6
1892 B	7 2 6 53.8	0.016	11 13.358	0 5.14	9 12.84	3 28.0	0 5.1
1893	6 13 46 22.4	0.015	0 13.723	0 5.36	8 12.66	1 28.7	2 11.6
1894	7 1 25 51.0	0.016	1 14.088	0 5.58	7 12.48	11 29.4	4 18.0
1895	7 13 5 19.5	0.018	2 14.454	0 5.80	6 12.33	10 0.1	6 21.5
1896 B	0 20 45 12.2	0.000	3 14.223	11 28.96	5 5.73	7 21.8	2 4.9
1897	7 12 24 16.6	0.018	4 15.184	0 6.25	4 12.08	6 1.3	11 7.4
1898	0 20 4 9.4	0.000	5 14.951	11 29.41	3 5.19	3 26.1	6 17.8
1899	1 7 43 37.9	0.001	6 15.319	11 29.63	2 5.37	1 26.7	8 24.2
1900	1 19 23 6.5	0.002	7 15.684	11 29.86	1 5.24	11 27.3	11 0.7

and the Arguments of the Inequalities.

YEARS.	6	7	8	9	I	II	III	IV
1880 B	1 20.3	2 12.8	8 0.0	7 7.5	1 0.45	4 22.47	5 5.5	0 12.1
1881	2 18.1	3 12.5	5 2.9	4 8.5	2 0.85	5 25.43	6 6.6	1 24.6
1882	3 15.9	4 12.2	2 5.7	1 9.4	3 1.28	6 28.41	7 7.8	3 7.1
1883	4 13.7	5 11.9	11 8.6	10 10.4	4 1.70	8 1.39	8 8.9	4 19.6
1884 B	5 11.5	6 11.6	8 11.5	7 11.4	5 2.09	9 4.34	9 10.0	6 2.1
1885	6 9.3	7 11.3	5 14.4	4 12.4	6 2.46	10 7.26	10 11.0	7 14.5
1886	7 7.0	8 11.0	2 17.3	1 19.4	7 2.81	11 10.17	11 12.0	8 26.9
1887	8 4.8	9 10.6	11 20.2	10 14.3	8 3.15	0 13.06	0 13.1	10 9.4
1888 B	9 2.5	10 10.2	8 23.1	7 15.3	9 3.49	1 15.96	1 14.1	11 21.8
1889	10 0.2	11 9.9	5 26.0	4 16.3	10 3.83	2 18.86	2 15.2	1 4.2
1890	10 28.0	0 9.5	2 28.8	1 17.3	11 4.19	3 21.78	3 16.2	2 16.7
1891	11 25.8	1 9.2	0 1.7	10 18.2	0 4.57	4 24.71	4 17.3	3 29.1
1892 B	0 23.6	2 8.9	9 4.6	7 19.2	1 4.97	5 27.07	5 18.4	5 11.6
1893	1 21.4	3 8.6	6 7.5	4 20.2	2 5.38	7 0.04	6 19.5	6 24.1
1894	2 19.2	4 8.3	3 10.1	1 21.2	3 5.78	8 3.59	7 20.6	8 6.6
1895	3 16.9	5 8.0	0 13.3	10 22.2	4 6.16	9 6.53	8 21.6	9 19.1
1896 B	4 14.1	6 7.1	9 10.8	7 17.8	5 5.92	10 8.80	9 22.1	11 0.7
1897	5 12.4	7 7.3	6 19.0	4 24.1	6 6.86	11 12.34	10 23.7	0 14.0
1898	6 9.6	8 6.4	3 16.6	1 19.8	7 6.61	0 14.60	11 24.2	1 25.5
1899	7 7.4	9 6.0	0 19.5	10 20.8	8 6.96	1 17.50	0 25.2	3 8.0
1900	8 5.1	10 5.7	9 22.3	7 21.7	9 7.31	2 20.41	1 26.2	4 20.4

TABLE III.   Perturbations of Jupiter and other Inequalities.

$$0.477759\,(J + \varphi + \delta E) + 493.2\,(\varphi_1 + \delta r) + (^*)\,2.823 \sin(5\tilde{u} - 2w_0 - 34.542) - (^*)\,5.775 \sin(H - A_{01}).$$

Years and tenths.	Perturb.	Diff.	Years and tenths.	Perturb.	Diff.	Years and tenths.	Perturb.	Diff.	Years and tenths.	Perturb.	Diff.
1880.0	3 20.2	+4.0	1883.0	8 16.2	+7.1	1886.0	8 8.3	−5.4	1889.0	5 16.3	−3.5
1	3 33.2	4.8	1	8 21.3	6.4	1	8 2.9	−5.7	1	5 12.8	−3.1
2	3 38.0	5.6	2	8 29.7	5.9	2	7 57.2	−5.7	2	5 9.7	−2.7
3	3 43.6	6.4	3	8 35.6	5.3	3	7 51.5	−5.9	3	5 7.0	−2.4
4	3 50.0	+7.1	4	8 40.9	+4.6	4	7 45.6	−6.0	4	5 4.6	−1.9
5	3 57.1	7.8	5	8 45.5	4.0	5	7 39.6	−6.0	5	5 2.7	−1.6
6	4 4.9	8.4	6	8 49.5	3.4	6	7 33.6	−6.2	6	5 1.1	−1.1
7	4 13.3	9.0	7	8 52.9	2.8	7	7 27.4	−6.3	7	5 0.0	−0.7
8	4 22.3	9.6	8	8 55.7	2.3	8	7 21.1	−6.2	8	4 59.3	−0.2
9	4 31.9	+10.2	9	8 58.0	+1.7	9	7 14.9	−6.4	9	4 59.1	+0.2
1881.0	4 42.1	10.7	1884.0	8 59.7	1.8	1887.0	7 8.5	−6.4	1890.0	4 59.3	0.7
1	4 52.8	11.0	1	9 0.8	0.7	1	7 2.1	−6.4	1	5 0.0	1.1
2	5 3.8	11.5	2	9 1.5	+0.2	2	6 55.7	−6.4	2	5 1.1	1.6
3	5 15.3	11.6	3	9 1.7	−0.3	3	6 49.3	−6.4	3	5 2.7	2.1
4	5 26.9	+11.9	4	9 1.4	−0.7	4	6 42.9	−6.4	4	5 4.8	+2.6
5	5 38.8	12.0	5	9 0.7	−1.2	5	6 36.5	−6.4	5	5 7.4	3.1
6	5 50.8	11.9	6	8 59.5	−1.8	6	6 30.1	−6.5	6	5 10.5	3.5
7	6 2.7	12.0	7	8 57.9	−2.0	7	6 23.8	−6.2	7	5 14.0	4.0
8	6 14.7	11.8	8	8 55.9	−2.4	8	6 17.6	−6.0	8	5 18.0	4.4
9	6 26.5	+11.8	9	8 53.5	−2.7	9	6 11.6	−6.0	9	5 22.4	+4.8
1882.0	6 38.3	11.5	1885.0	8 50.8	−3.1	1888.0	6 5.6	−5.8	1891.0	5 27.2	5.4
1	6 49.8	11.3	1	8 47.7	−3.3	1	5 59.8	−5.6	1	5 32.6	5.8
2	7 1.1	10.9	2	8 44.4	−3.7	2	5 54.2	−5.5	2	5 38.4	6.1
3	7 12.0	10.6	3	8 40.7	−3.9	3	5 48.7	−5.4	3	5 44.5	6.5
4	7 22.6	+10.2	4	8 36.8	−4.2	4	5 43.3	−5.1	4	5 51.0	+6.9
5	7 32.8	9.7	5	8 32.6	−4.4	5	5 38.2	−4.9	5	5 57.9	7.2
6	7 42.5	9.3	6	8 28.2	−4.6	6	5 33.3	−4.7	6	6 5.1	7.5
7	7 51.8	8.7	7	8 23.6	−4.9	7	5 28.6	−4.4	7	6 12.6	7.8
8	8 0.5	8.1	8	8 18.7	−5.1	8	5 24.2	−4.1	8	6 20.4	8.0
9	8 8.6	+7.6	9	8 13.6	−5.3	9	5 20.1	−3.8	9	6 28.4	+8.2
1883.0	8 16.2		1886.0	8 8.3		1889.0	5 16.3		1892.0	6 36.6	

TABLE III.   Perturbations of Jupiter and other Inequalities.

$$0.477759 \,(J + \rho + \delta E) + 493.2 \,(\delta_1 + \delta \tau) + (^*)\, 2.823 \sin (5\,\bar{u} - 2\,u_0 - 34.542) - (^*)\, 5.775 \sin (\Pi - \bar{u}'\,\text{m}).$$

Years and tenths.	Perturb.	Diff.	Years and tenths.	Perturb.	Diff.	Years and tenths.	Perturb.	Diff.	Years and tenths.	Perturb.	Diff.
	m s	s		m s	s		m s	s		m s	s
1892.0	6 36.6	+8.4	1894.0	9 14.2	+4.9	1896.0	9 23.9	-3.8	1898.0	7 38.6	-5.5
1	6 45.0	8.6	1	9 19.1	4.5	1	9 20.1	-4.1	1	7 33.1	-5.6
2	6 53.6	8.7	2	9 23.6	4.1	2	9 16.0	-4.3	2	7 27.5	-5.4
3	7 2.3	8.8	3	9 27.7	3.5	3	9 11.7	-4.6	3	7 22.1	-5.3
4	7 11.1	+8.8	4	9 31.2	+3.1	4	9 7.1	-4.8	4	7 16.8	-5.2
5	7 19.9	8.9	5	9 34.3	2.6	5	9 2.3	-5.1	5	7 11.6	-5.1
6	7 28.8	8.9	6	9 36.9	2.1	6	8 57.2	-5.1	6	7 6.5	-4.9
7	7 37.7	8.8	7	9 39.0	1.5	7	8 52.1	-5.3	7	7 1.6	-4.8
8	7 46.5	8.7	8	9 40.5	1.1	8	8 46.8	-5.5	8	6 56.8	-4.6
9	7 55.2	+8.6	9	9 41.6	+0.6	9	8 41.3	-5.4	9	6 52.2	-4.4
1893.0	8 3.8	8.4	1895.0	9 42.2	+0.1	1897.0	8 35.9	-5.6	1899.0	6 47.8	-4.3
1	8 12.2	8.2	1	9 42.3	-0.4	1	8 30.3	-5.7	1	6 43.5	-4.0
2	8 20.4	7.9	2	9 41.9	-0.8	2	8 24.6	-5.7	2	6 39.5	-3.9
3	8 28.3	7.7	3	9 41.1	-1.3	3	8 18.9	-5.7	3	6 35.6	-3.5
4	8 36.0	+7.3	4	9 39.8	-1.7	4	8 13.2	-5.8	4	6 32.1	-3.4
5	8 43.3	6.9	5	9 38.1	-2.1	5	8 7.4	-5.8	5	6 28.7	-3.1
6	8 50.2	6.6	6	9 36.0	-2.5	6	8 1.6	-5.8	6	6 25.6	-2.8
7	8 56.8	6.2	7	9 33.5	-2.9	7	7 55.8	-5.7	7	6 22.8	-2.5
8	9 3.0	5.8	8	9 30.6	-3.2	8	7 50.1	-5.8	8	6 20.3	-2.1
9	9 8.8	+5.4	9	9 27.4	-3.5	9	7 44.3	-5.7	9	6 18.2	-1.9
1894.0	9 14.2		1896.0	9 23.9		1898.0	7 38.6		1900.0	6 16.3	

TABLE I.   Epochs of the Mean Conjunctions

YEARS.	MEAN CONJUNCTIONS.		1	2	3	4
	Days and parts of a day, Paris mean civil time.	Fraction of year.				
	Jan. h m s		s. o	s. o	s. o	s. o
1880 B	0 20 29 8.9	0.002	11 8.706	11 29.29	9 11.75	6 10.9
1881	3 10 21 41.3	0.007	0 9.329	0 2.56	8 14.37	11 16.3
1882	7 0 14 13.7	0.016	1 9.951	0 5.83	7 16.96	4 21.7
1883	10 14 6 46.2	0.026	2 10.573	0 9.10	6 19.55	9 27.1
1884 B	14 3 59 18.6	0.036	3 11.196	0 12.37	5 22.18	3 2.5
1885	16 17 51 51.0	0.043	4 11.818	0 15.64	4 24.83	8 7.8
1886	3 13 39 16.5	0.007	5 11.048	0 2.40	3 12.37	9 11.6
1887	7 3 31 48.9	0.017	6 11.670	0 5.67	2 15.05	2 16.9
1888 B	10 17 24 21.3	0.027	7 12.293	0 8.94	1 17.73	7 22.2
1889	13 7 16 53.8	0.034	8 12.915	0 12.21	0 20.40	0 27.5
1890	0 3 4 19.3	0.000	9 12.145	11 28.97	11 7.94	2 1.2
1891	3 16 56 51.7	0.007	10 12.768	0 2.24	10 10.57	7 6.6
1892 B	7 6 49 24.1	0.017	11 13.390	0 5.51	9 13.19	0 12.0
1893	9 20 41 56.5	0.024	0 14.012	0 8.78	8 15.79	5 17.4
1894	13 10 34 29.0	0.034	1 14.634	0 12.05	7 18.41	10 22.8
1895	0 6 21 54.4	0.000	2 13.864	11 28.81	6 5.93	11 26.5
1896 B	3 20 14 26.9	0.008	3 14.486	0 2.08	5 8.58	5 1.8
1897	6 10 6 59.3	0.015	4 15.109	0 5.35	4 11.26	10 7.2
1898	9 23 59 31.7	0.025	5 15.731	0 8.62	3 13.93	3 12.5
1899	13 13 52 4.1	0.034	6 16.353	0 11.90	2 16.60	8 17.8
1900	0 9 39 29.6	0.000	7 15.583	11 28.65	1 4.14	9 21.5

and the Arguments of the Inequalities.

YEARS.	5	6.	7	I	II	III	IV
1880 B	4 9.2	2 12.50	1 20.1	0 29.50	5 4.48	4 21.4	0 11.0
1881	0 23.2	3 12.45	2 18.1	2 0.16	6 5.84	5 24.6	1 23.8
1882	9 7.3	4 12.43	3 16.2	3 0.85	7 7.22	6 27.9	3 6.7
1883	5 21.4	5 12.41	4 14.2	4 1.53	8 8.60	8 1.1	4 19.6
1884 B	2 5.4	6 12.35	5 12.3	5 2.16	9 9.95	9 4.4	6 2.4
1885	10 19.2	7 12.27	6 10.3	6 2.80	10 11.27	10 7.6	7 15.2
1886	10 16.0	8 10.82	7 7.0	7 2.02	11 11.15	11 9.2	8 26.1
1887	6 29.8	9 10.71	8 4.9	8 2.62	0 12.44	0 12.4	10 8.8
1888 B	3 13.6	10 10.60	9 2.9	9 3.21	1 13.74	1 15.6	11 21.6
1889	11 27.4	11 10.50	10 0.9	10 3.82	2 15.04	2 18.8	1 4.4
1890	11 24.2	0 9.05	10 27.6	11 3.04	3 14.93	3 20.5	2 15.3
1891	8 8.2	1 8.99	11 25.6	0 3.68	4 16.27	4 23.7	3 28.1
1892 B	4 22.2	2 8.94	0 23.6	1 4.34	5 17.63	5 26.9	5 11.0
1893	1 6.2	3 8.91	1 21.7	2 5.01	6 18.99	7 0.2	6 23.8
1894	9 20.2	4 8.86	2 19.7	3 5.67	7 20.35	8 3.4	8 6.7
1895	9 17.2	5 7.43	3 16.4	4 4.91	8 20.26	9 5.1	9 17.5
1896 B	6 1.0	6 7.34	4 14.1	5 5.53	9 21.57	10 8.3	11 0.4
1897	2 14.8	7 7.24	5 12.4	6 6.13	10 22.87	11 11.5	0 13.1
1898	10 28.6	8 7.14	6 10.4	7 6.73	11 24.17	0 14.6	1 25.9
1899	7 12.4	9 7.01	7 8.4	8 7.34	0 25.47	1 17.8	3 8.7
1900	7 9.2	10 5.59	8 5.1	9 6.56	1 25.36	2 19.5	4 59.6

TABLE III.   Perturbations of Jupiter and other Inequalities.

$$1.1169035 \left(J + \phi + \delta E \right) + 495.2 \left(\phi_1 + \delta r \right) + (\text{*}) \, 15.581 \sin \left(5\,\bar{u} - 2\,u_o - 34.542 \right) + 16.694 \sin \left(\text{II} - \Lambda_{\text{IV}} \right).$$

Years and tenths.	Perturb.	Diff.	Years and tenths.	Perturb.	Diff.	Years and tenths.	Perturb.	Diff.	Years and tenths.	Perturb.	Diff.
	m s	s		m s	s		m s	s		m s	s
1880.0	8 45.4	+9.5	1883.0	19 56.5	+16.3	1886.0	19 32.5	−12.9	1889.0	12 49.6	−8.0
1	8 54.9	11.5	1	20 12.8	14.9	1	19 19.6	−13.3	1	12 41.6	−7.2
2	9 6.4	13.3	2	20 27.7	13.5	2	19 6.3	−13.6	2	12 34.4	−6.4
3	9 19.7	14.9	3	20 41.2	12.0	3	18 52.7	−13.9	3	12 28.0	−5.4
4	9 34.6		4	20 53.2		4	18 38.8		4	12 22.6	
		+16.8			+10.7			−14.1			−4.5
5	9 51.4	18.3	5	21 3.9	9.1	5	18 24.7	−14.3	5	12 18.1	−3.6
6	10 9.7	19.8	6	21 13.0	7.8	6	18 10.4	−14.5	6	12 14.5	−2.5
7	10 29.5	21.2	7	21 20.8	6.4	7	17 55.9	−14.6	7	12 12.0	−1.5
8	10 50.7	22.6	8	21 27.2	5.1	8	17 41.3	−14.7	8	12 10.5	−0.5
9	11 13.3		9	21 32.3		9	17 26.6		9	12 10.0	
		+23.8			+3.7			−14.9			+0.6
1881.0	11 37.1	25.1	1884.0	21 36.0	2.6	1887.0	17 11.7	−15.0	1890.0	12 10.6	1.7
1	12 2.2	26.0	1	21 38.6	1.4	1	16 56.7	−15.0	1	12 12.3	2.8
2	12 28.2	26.7	2	21 40.0	+0.2	2	16 41.7	−15.1	2	12 15.1	3.8
3	12 54.9	27.4	3	21 40.2	−0.9	3	16 26.6	−15.0	3	12 18.9	5.0
4	13 22.3		4	21 39.3		4	16 11.6		4	12 23.9	
		+27.7			−2.0			−15.0			+6.1
5	13 50.0	27.9	5	21 37.3	−3.0	5	15 56.6	−14.8	5	12 30.0	7.2
6	14 17.9	28.0	6	21 34.3	−3.9	6	15 41.8	−14.7	6	12 37.2	8.3
7	14 45.9	27.9	7	21 30.4	−4.9	7	15 27.1	−14.4	7	12 45.5	9.4
8	15 13.8	27.7	8	21 25.5	−5.7	8	15 12.7	−14.2	8	12 54.9	10.4
9	15 41.5		9	21 19.8		9	14 58.5		9	13 5.3	
		+27.3			−6.6			−14.0			+11.5
1882.0	16 8.8	26.9	1885.0	21 13.2	−7.2	1888.0	14 44.5	−13.5	1891.0	13 16.8	12.5
1	16 35.7	26.3	1	21 6.0	−8.0	1	14 31.0	−13.3	1	13 29.3	13.5
2	17 2.0	25.5	2	20 58.0	−8.7	2	14 17.7	−12.8	2	13 42.8	14.4
3	17 27.5	24.7	3	20 49.3	−9.4	3	14 4.9	−12.4	3	13 57.2	15.3
4	17 52.2		4	20 39.9		4	13 52.5		4	14 12.5	
		+23.6			−9.9			−11.9			+16.1
5	18 15.8	22.6	5	20 30.0	−10.5	5	13 40.6	−11.5	5	14 28.6	16.8
6	18 38.4	21.4	6	20 19.5	−11.0	6	13 29.1	−10.8	6	14 45.4	17.6
7	18 59.8	20.2	7	20 8.5	−11.6	7	13 18.3	−10.3	7	15 3.0	18.1
8	19 20.0	18.9	8	19 56.9	−12.0	8	13 8.0	−9.5	8	15 21.1	18.7
9	19 38.9		9	19 44.9		9	12 58.5		9	15 39.8	
		+17.6			−12.4			−8.9			+19.2
1883.0	19 56.5		1886.0	19 32.5		1889.0	12 49.6		1892.0	15 59.0	

TABLE III.   Perturbations of Jupiter and other Inequalities.

$$1.1160035 \, (J + \phi + \delta \, E) + 493.2 \, (\lambda_1 + \delta \, \tau) + (') \, 15.581 \sin (5 \, \bar{u} - 2 \, u_0 - 34.542) + 16.694 \sin (\Pi - / \, IV).$$

Years and tenths.	Perturb.	Diff.	Years and tenths.	Perturb.	Diff.	Years and tenths.	Perturb.	Diff.	Years and tenths.	Perturb.	Diff.	Years and tenths.	Perturb.	Diff.
	m s	s		m s	s		m s	s		m s	s		m s	s
1892.0	15 59.0	+19.6	1894.0	22 5.0	+11.4	1896.0	22 24.2	− 9.1	1898.0	18 15.7	−13.2			
1	16 18.6	20.0	1	22 16.4	10.4	1	22 15.1	− 9.7	1	18 2.5	−13.0			
2	16 38.6	20.3	2	22 26.8	9.2	2	22 5.4	−10.3	2	17 49.5	−12.7			
3	16 58.9	20.5	3	22 36.0	8.2	3	21 55.1	−10.9	3	17 36.8	−12.5			
4	17 19.4	+20.6	4	22 44.2	+ 7.0	4	21 44.2	−11.3	4	17 24.3	−12.1			
5	17 40.0	20.7	5	22 51.2	5.9	5	21 32.9	−11.8	5	17 12.2	−11.9			
6	18 0.7	20.7	6	22 57.1	4.6	6	21 21.1	−12.2	6	17 0.3	−11.6			
7	18 21.4	20.5	7	23 1.7	3.5	7	21 8.9	−12.5	7	16 48.7	−11.2			
8	18 41.9	20.2	8	23 5.2	2.3	8	20 56.4	−12.8	8	16 37.5	−11.2			
9	19 2.1	+19.9	9	23 7.5	+ 1.2	9	20 43.6	−13.1	9	16 26.7	−10.8			
1893.0	19 22.0	19.6	1895.0	23 8.7	0.0	1897.0	20 30.5	−13.2	1899.0	16 16.3	− 9.9			
1	19 41.6	18.9	1	23 8.7	− 1.1	1	20 17.3	−13.3	1	16 6.4	− 9.4			
2	20 0.5	18.4	2	23 7.6	− 2.1	2	20 4.0	−13.5	2	15 57.0	− 9.0			
3	20 18.9	17.7	3	23 5.5	− 3.6	3	19 50.5	−13.5	3	15 48.0	− 8.3			
4	20 36.6	+17.0	4	23 1.9	− 3.8	4	19 37.0	−13.6	4	15 39.7	− 7.8			
5	20 53.6	16.1	5,	22 58.1	− 5.1	5	19 23.4	−13.6	5	15 31.9	− 7.2			
6	21 9.7	15.2	6	'22 53.0	− 6.0	6	19 9.8	−13.6	6	15 24.7	− 6.5			
7	21 24.9	14.4	7	22 47.0	− 6.8	7	18 56.2	−13.6	7	15 18.2	− 5.8			
8	21 39.3	13.3	8	22 40.2	− 7.7	8	18 42.6	−13.5	8	15 12.4	− 5.1			
9	21 52.6	+12.4	9	22 32.5	− 8.3	9	18 29.1	−13.4	9	15 7.3	− 4.3			
1894.0	22 5.0		1896.0	22 24.2		1898.0	18 15.7		1900.0	15 3.0				

TABLE A.

LONGITUDES OF OBSERVATORIES.

West longitudes, positive.

Place of Observatory.	Longitude from Paris.	Place of Observatory.	Longitude from Paris.
	h m s		h m s
Åbo	− 1 19 47.3	Liverpool	+ 0 21 21.1
Albany	+ 5 4 20.3	Madras	− 5 11 36.2
Allegheny	+ 5 29 23.8	Madrid	+ 0 24 6.4
Altona	− 0 30 25.1	Mannheim	− 0 21 30.0
Ann Arbor	+ 5 44 16.2	Markree	+ 0 43 9.4
Armagh	+ 0 35 56.5	Marseilles	− 0 12 7.5
Athens	− 1 25 34.1	Milan	− 0 27 25.2
Berlin	− 0 44 14.3	Modena	− 0 34 22.5
Bilk	− 0 17 44.3	Moscow	− 2 20 55.8
Bonn	− 0 19 3.0	Munich	− 0 37 5.0
Breslau	− 0 58 48.7	Naples	− 0 47 37.9
Brussels	− 0 8 7.8	New York	+ 5 5 17.7
Cambridge (Eng.)	+ 0 8 58.4	Nicolajew	− 1 58 34.1
Cambridge (Mass.)	+ 4 53 52.0	Olmuetz	− 0 59 42.4
Cape of Good Hope	− 1 4 34.6	Oxford	+ 0 14 23.7
Chicago	+ 5 59 47.8	Padua	− 0 38 8.2
Christiania	− 0 33 33.2	Palermo	− 0 44 4.0
Cincinnati	+ 5 47 20.1	Parammatta	− 9 51 45.2
Clinton	+ 5 10 58.5	Philadelphia	+ 5 9 59.5
Copenhagen	− 0 40 57.6	Prague	− 0 48 20.5
Cordoba	+ 4 26 6.0	Pulkowa	− 1 51 57.6
Cracow	− 1 10 29.8	Rome	− 0 40 35.1
Dorpat	− 1 37 33.0	San Fernando	+ 0 34 10.6
Dublin	+ 0 34 43.1	Santiago	+ 4 52 3.4
Durham	+ 0 15 40.8	Seufienburg	− 0 56 29.6
Edinburgh	+ 0 22 4.1	Speyer	− 0 24 25.0
Florence	− 0 35 42.1	Stockholm	− 1 2 53.3
Geneva	− 0 15 16.2	St. Petersburg	− 1 51 52.4
Georgetown	+ 5 17 39.4	Upsala	− 1 1 9.7
Goettingen	− 0 30 25.5	Utrecht	− 0 11 10.6
Gotha	− 0 33 29.9	Vienna	− 0 56 11.1
Greenwich	+ 0 9 21.1	Washington	+ 5 17 33.2
Hamburg	− 0 30 32.5	Wilna	− 1 31 50.3
Helsingfors	− 1 30 28.3		
Hudson	+ 5 35 5.2		
Kasan	− 3 7 7.7		
Koenigsberg	− 1 12 32.1		
Kremsmuenster	− 0 47 12.0		
Leipsic	− 0 40 13.4		
Leyden	− 0 8 35.1		

TABLES

FOR

FINDING THE CONFIGURATIONS

OF THE

SATELLITES OF JUPITER.

5

TABLE I. Epochs of the Mean Longitude, and the Arguments of the Inequalities,

for January 1, Paris mean midnight.

YEARS.	Mean Longitude.	1		2		3		4		5	
1880 B	6 9.80	9	19.2	11	8.4	0	0.8	2	5.9	7	21.6
1881	4 26.77	8	19.5	0	8.8	0	1.6	11	29.7	6	11.5
1882	8 20.25	7	18.9	1	9.1	0	1.3	6	11.4	10	5.0
1883	0 13.73	6	18.4	2	9.5	0	1.1	0	23.1	1	28.5
1884 B	1 7.21	5	17.8	3	9.8	0	0.8	7	4.8	5	22.0
1885	2 24.18	4	18.1	4	10.2	0	1.6	1	28.6	4	8.9
1886	6 17.67	3	17.5	5	10.5	0	1.3	11	10.3	8	2.4
1887	10 11.15	2	16.9	6	10.8	0	1.1	5	22.0	11	25.9
1888 B	2 4.63	1	16.4	7	11.2	0	0.8	0	3.7	3	19.4
1889	0 21.61	0	16.7	8	11.6	0	1.6	9	27.5	2	6.3
1890	4 15.09	11	16.1	9	11.9	0	1.3	4	9.2	5	20.8
1891	8 8.57	10	15.5	10	12.2	0	1.1	10	20.9	9	23.2
1892 B	0 2.05	9	14.9	11	12.6	0	0.8	5	2.6	1	16.7
1893	10 19.03	8	15.3	0	13.0	0	1.6	2	26.4	0	3.6
1894	2 12.51	7	14.7	1	13.3	0	1.3	9	8.1	3	27.1
1895	6 5.99	6	14.1	2	13.6	0	1.1	3	19.8	7	20.6
1896 B	9 29.47	5	13.5	3	14.0	0	0.8	10	1.5	11	14.1
1897	8 16.45	4	13.8	4	14.4	0	1.5	7	25.3	10	1.0
1898	0 9.93	3	13.3	5	14.7	0	1.2	2	7.0	1	24.5
1899	4 3.41	2	12.7	6	15.0	0	1.0	8	18.7	5	17.9
1900	7 26.89	1	12.1	7	15.3	0	0.8	3	0.4	9	11.4

TABLE I. Epochs of the Mean Longitude, and the Arguments of the Inequalities,

for January 1, Paris mean midnight.

YEARS.	Mean Longitude.	1	2	3	4	5	6	7
1880 B	4 3.87	9 19.2	11 8.4	0 0.8	10 2.9	5 18.7	5 0.2	9 10.6
1881	4 27.04	8 19.5	0 8.8	0 1.6	8 29.8	6 11.8	6 5.4	10 6.3
1882	2 6.82	7 18.9	1 9.1	0 1.3	6 5.7	3 23.6	3 29.3	7 20.6
1883	11 20.61	6 18.4	2 9.5	0 1.1	3 11.5	1 5.4	1 23.1	5 4.9
1884 B	9 2.40	5 17.8	3 9.8	0 0.8	0 17.4	10 17.2	11 17.0	2 19.3
1885	9 25.56	4 18.1	4 10.2	0 1.6	11 14.3	11 10.3	0 22.2	3 15.0
1886	7 7.35	3 17.5	5 10.5	0 1.3	8 20.1	8 22.1	10 16.0	0 29.3
1887	4 19.14	2 16.9	6 10.8	0 1.1	5 26.0	6 3.9	8 9.9	10 13.6
1888 B	2 0.93	1 16.4	7 11.2	0 0.8	3 1.8	3 15.6	6 3.7	7 27.9
1889	2 24.09	0 16.7	8 11.6	0 1.6	1 28.7	4 8.8	7 9.0	8 23.7
1890	0 5.88	11 16.1	9 11.9	0 1.3	11 4.6	1 20.6	5 2.8	6 8.0
1891	9 17.67	10 15.5	10 12.2	0 1.1	8 10.4	11 2.3	2 26.7	3 22.3
1892 B	6 29.45	9 14.9	11 12.6	0 0.8	5 16.3	8 14.1	0 20.5	1 6.6
1893	7 22.62	8 15.3	0 13.0	0 1.6	4 13.2	9 7.3	1 25.8	2 2.3
1894	5 4.41	7 14.7	1 13.3	0 1.3	1 19.0	6 19.0	11 19.6	11 16.7
1895	2 16.19	6 14.1	2 13.6	0 1.1	10 24.9	4 0.8	9 13.5	9 1.0
1896 B	11 27.98	5 13.5	3 14.0	0 0.8	8 0.7	1 12.6	7 7.3	6 15.3
1897	0 21.14	4 13.8	4 14.4	0 1.5	6 27.6	2 5.7	8 12.6	7 11.0
1898	10 2.93	3 13.3	5 14.7	0 1.2	4 3.5	11 17.5	6 6.4	4 25.4
1899	7 14.72	2 12.7	6 15.0	0 1.0	1 9.3	8 29.3	4 0.2	2 9.7
1900	4 26.51	1 12.1	7 15.3	0 0.8	10 15.2	6 11.0	1 24.1	11 24.0

TABLE I. Epochs of the Mean Longitude, and the Arguments of the Inequalities,

for January 1, Paris mean midnight.

YEARS.	Mean Longitude.	1	2	3	4	5	6	7	8	9
1880 B	6 0.92	9 19.2	11 8.4	0 0.8	10 2.9	8 5.5	8 28.0	7 15.8	11 7.6	11 20.7
1881	7 27.18	8 19.5	0 8.8	0 1.6	8 29.8	9 29.1	10 23.6	9 12.0	1 6.4	1 17.7
1882	8 3.12	7 18.9	1 9.1	0 1.3	6 5.7	10 2.5	10 28.8	9 18.0	1 14.9	1 24.3
1883	8 9.06	6 18.4	2 9.5	0 1.1	3 11.5	10 5.8	11 4.0	9 23.9	1 23.4	2 0.9
1884 B	8 15.00	5 17.8	3 9.8	0 0.8	0 17.4	10 9.1	11 9.3	9 29.9	2 1.9	2 7.6
1885	10 11.26	4 18.1	4 10.2	0 1.6	11 14.3	0 2.7	1 4.8	11 26.0	4 0.7	4 4.5
1886	10 17.20	3 17.5	5 10.5	0 1.3	8 20.1	0 6.1	1 10.1	0 2.0	4 9.2	4 11.1
1887	10 23.14	2 16.9	6 10.8	0 1.1	5 26.0	0 9.4	1 15.3	0 8.0	4 17.6	4 17.7
1888 B	10 29.08	1 16.4	7 11.2	0 0.8	3 1.8	0 12.7	1 20.5	0 13.9	4 26.1	4 24.3
1889	0 25.34	0 16.7	8 11.6	0 1.6	1 28.7	2 6.3	3 16.1	2 10.1	6 24.9	6 21.3
1890	1 1.28	11 16.1	9 11.9	0 1.3	11 4.6	2 9.7	3 21.3	2 16.0	7 3.4	6 27.9
1891	1 7.22	10 15.5	10 12.2	0 1.1	8 10.4	2 13.0	3 26.5	2 22.0	7 11.9	7 4.5
1892 B	1 13.16	9 14.9	11 12.6	0 0.8	5 16.3	2 16.3	4 1.7	2 27.8	7 20.4	7 11.1
1893	3 9.42	8 15.3	0 13.0	0 1.6	4 13.2	4 10.0	5 27.3	4 24.2	9 19.2	9 8.1
1894	3 15.36	7 14.7	1 13.3	0 1.3	1 19.0	4 13.3	6 2.5	5 0.0	9 27.6	9 14.7
1895	3 21.30	6 14.1	2 13.6	0 1.1	10 24.9	4 16.6	6 7.8	5 6.0	10 6.1	9 21.3
1896 B	3 27.24	5 13.5	3 14.0	0 0.8	8 0.7	4 19.9	6 13.0	5 11.9	10 14.6	9 27.9
1897	5 23.50	4 13.8	4 14.4	0 1.5	6 27.6	6 13.6	8 8.5	7 8.2	0 13.4	11 24.9
1898	5 29.44	3 13.3	5 14.7	0 1.2	4 3.5	6 16.9	8 13.7	7 14.0	0 21.9	0 1.5
1899	6 5.38	2 12.7	6 15.0	0 1.0	1 9.3	6 20.2	8 18.9	7 20.0	1 0.4	0 8.1
1900	6 11.32	1 12.1	7 15.3	0 0.8	10 15.2	6 23.5	8 24.2	7 25.9	1 8.8	0 14.7

TABLE I. Epochs of the Mean Longitude, and the Arguments of the Inequalities,

for January 1, Paris mean midnight.

YEARS.	Mean Longitude.	1	2	3	4	5	6	7
		s. °	s. °	s. °	s. °	s. °	s. °	s. °	s. °
1880 B	11 17.40	9 19.2	11 8.4	0 0.8	2 14.4	1 2.2	5 7.2	4 24.1
1881	10 22.42	8 19.5	0 8.8	0 1.6	1 18.7	0 7.2	4 12.9	4 1.6
1882	9 5.88	7 18.9	1 9.1	0 1.3	0 1.5	10 20.7	2 27.0	2 17.6
1883	7 19.33	6 18.4	2 9.5	0 1.1	10 14.2	9 4.1	1 11.2	1 3.6
1884 B	6 2.79	5 17.8	3 9.8	0 0.8	8 27.0	7 17.6	11 25.3	11 19.6
1885	5 7.81	4 18.1	4 10.2	0 1.6	8 1.3	6 22.6	11 1.0	10 27.2
1886	3 21.27	3 17.5	5 10.5	0 1.3	6 14.0	5 6.0	9 15.1	9 13.2
1887	2 4.72	2 16.9	6 10.8	0 1.1	4 26.8	3 19.4	7 29.3	7 29.2
1888 B	0 18.18	1 16.4	7 11.2	0 0.8	3 9.5	2 2.9	6 13.4	6 15.2
1889	11 23.20	0 16.7	8 11.6	0 1.6	2 13.8	1 7.9	5 19.1	5 22.7
1890	10 6.66	11 16.1	9 11.9	0 1.3	0 26.5	11 21.3	4 3.2	4 8.7
1891	8 20.11	10 15.5	10 12.2	0 1.1	11 9.3	10 4.8	2 17.4	2 24.7
1892 B	7 3.57	9 14.9	11 12.6	0 0.8	9 22.0	8 18.2	1 1.5	1 10.7
1893	6 8.60	8 15.3	0 13.0	0 1.6	8 26.4	7 23.2	0 7.2	0 18.3
1894	4 22.05	7 14.7	1 13.3	0 1.3	7 9.1	6 6.7	10 21.3	11 4.3
1895	3 5.50	6 14.1	2 13.6	0 1.1	5 21.8	4 20.1	9 5.5	9 20.3
1896 B	1 18.96	5 13.5	3 14.0	0 0.8	4 4.6	3 3.6	7 19.6	8 6.3
1897	0 23.99	4 13.8	4 14.4	0 1.5	3 8.9	2 8.6	6 25.3	7 13.8
1898	11 7.44	3 13.3	5 14.7	0 1.2	1 21.6	0 22.0	5 9.4	5 29.8
1899	9 20.90	2 12.7	6 15.0	0 1.0	0 4.4	11 5.4	3 23.6	4 15.8
1900	8 4.35	1 12.1	7 15.3	0 0.8	10 17.1	9 18.9	2 7.7	3 1.8

CORRECTIONS TO THE

TABLES ÉCLIPTIQUES DES SATELLITES DE JUPITER, ETC.,

PAR LE BARON DE DAMOISEAU, PARIS, 1836.

———————

Page.			
(III),	tenth line from the bottom.	*for* $u^1 - 2u_{II} + \pi_{III}$	*read* $u_1 - 2u_{II} + \pi_{III}$
(IV),	second line.	*for* I, II, III, IV, relativement	*read* I. II, III, relativement
(V),	$u_{II} - u_{III}$,	*for* 11. 27,7519	*read* 10. 27,7519
(VI),	Table X, third term.	*for* $+0''2,18\cos(\mathbf{1})\cos 3[\mathbf{3}-E+E']$	*read* $+0'',218\cos(\mathbf{1})\cos 3[\mathbf{3}-E+E']$
(VII),	eighteenth line,	*for* $u_{III} - \pi_{III} + 1,0015\,\varphi$	*read* $u_{III} - \pi_{III} + 1,0016\,\varphi$
(VII),	eleventh line from the bottom,	*for* $+i.79\,9^h\,59'$	*read* $+i.7!\,3^h\,59'$
(VII),	Table XII,	*for* $-0'',115\cos 2\,(\mathbf{2})$	*read* $-0'',115\cos\,(\mathbf{2})$
(VIII),	Table XXV,	*for* $-0,008079\,.\sin(III+1,0026\,E)$	*read* $-0,008079\,.\sin(III+1,0016\,E)$
(VIII),	Tables XXVIII—XXXII,	*for* $N=2P-P$	*read* $N=2P-P^{-1}$
(VIII),	$U - u_o$,	*for* 9. 6, 1550	*read* 9. 8,1550
(IX),	Table XII, third term,	*for* $+0''2,18\cos(\mathbf{1})\cos 3[\mathbf{3}-E+E']$	*read* $+0'',218\cos(\mathbf{1})\cos 3[\mathbf{3}-E+E']$
(X),	twelfth line,	*for* XIX	*read* XXIX
(XIV),	ninth line from the bottom,	*for* les tables XXIII—XXVI	*read* les tables XXIII—XXVII
(XIV),	sixth line from the bottom,	*for* la table XXVII	*read* la table XXVIII
(XVII),	nineteenth line,	*for* $2I + VI$	*read* $2I + \mathbf{6}$
2,	Conjonctions Moyennes, 1755,	*for* 1 10. 26. 28,1	*read* 1 10. 26. 8,1
6,	Conjonctions Moyennes, 1847,	*for* 2 0. 8. 27,1	*read* 2 0. 8. 27,0
7,	Arg. **5**, 1848,	*for* 3. 24,6	*read* 3. 25,6
9,	Arg. **5**, Janvier 14,	*for* 0. 6,7	*read* 0. 8,7
10,	Arg. **4**, Mars 6,	*for* 6. 26,08	*read* 6. 26,93
10,	Arg. **3**, Avril 23,	*for* 3. 12,93	*read* 3. 12,23
11,	Arg. **8**, Février 6,	*for* $0^s\,27^\circ,0$	*read* $0^s\,27^\circ,8$
12,	Arg. **1**, Mai 5,	*for* 2. 10,441	*read* 0. 10,441
12,	Arg. **1**, Mai 25,	*for* 1. 12,058	*read* 0. 12,058
14,	Arg. **1**, Juillet 29,	*for* 0. 17,490	*read* 0. 17,499
15,	Arg. **9**, Septemb. 22,	*for* 6. 16,3	*read* 6. 16,8
15,	Arg. 1, Septemb. 22,	*for* 0. 22,08	*read* 0. 22,06
16,	Révolutions, Octobre 27,	*for* 21. 1. 51,7	*read* 21. 1. 50,7
17,	Arg. **7**, Novemb. 25,	*for* 1. 26,7	*read* 0. 26,7
25,	Diff. 1847,2 to 1847,3,	*for* 0,6	*read* 0,7
26,	Perturb. 1860,1,	*for* 2. 27,9	*read* 2. 17,9
32,	$I^s\,4^\circ$,	*for* 2,01	*read* 3,01
32,	$X^s\,16^\circ$,	*for* 0,55	*read* 0,57
33,	V^s Equation 22°,	*for* 0. 46. 15,6	*read* 0. 46. 13,6
35,	I^s Diff. 28° to 29°,	*for* 6,5	*read* 6,6
35,	II^s Diff. 26° to 30°,	*for* 8,3 8,3 8,2 8,1	*read* 8,4 8,4 8,4 8,4
35,	III^s Diff. 17° to 18°,	*for* 3,5	*read* 8,5
36,	Arg. **1**, III^s 20°, Arg. **3**, XI^s 0° to 30°,	*for* 1,5 1,5 1,5 1,4	*read* 1,6 1,6 1,6 1,6
37,	Arg. **1**, VI^s 0°, Arg. **3**, XI^s 10°,	*for* 2,0	*read* 2,1
38,	Arg. **2**, II^s 0°, Arg. **3**, IX^s 20°,	*for* 6,6	*read* 6,2
39,	top of page,	*for* Suite de la TABLE XI	*read* Suite de la TABLE XII
39,	Arg. **2**, XI^s 10°, Arg. **3**, I^s 0°,	*for* 1,8	*read* 1,3

Page.		for	read
41,	top of page,	Suite de la TABLE V	Suite de la TABLE XIII
42,	heading of second column,	Argument **3**	Argument **5**
43,	I* Réduct. 23°,	0,7	0,9
44,	III* Nombre 20°,	0,5700	0,6700
45,	Arg. 0,4000, P,	1000	1,00
53,	Arg. **7**, 1834,	10. 0,9	10. 5,9
58,	Révolutions, Novemb. 15,	20. 50. 36,3	20. 50. 36,2
72,	V* Équation 27°,	1. 24. 36,6	1. 24. 26,6
73,	IX* Diff. 20° to 21°,	23,7	22,7
75,	Arg. **1**, 0* 20°, Arg. **3**, VIII* 0°,	2,1	2,2
75,	Arg. **1**, V* 10°, Arg. **3**, IV* 10°,	0,3	0,6
75,	Arg. **1**, V* 10°, Arg. **3**, IV* 20°,	1,6	1,0
76,	Arg. **1**, X* 20°, Arg. **3**, IX* 20°,	1,9	1,4
76,	Arg. **1**, XI* 10°, Arg. **3**, X* 0°,	1,5	1,3
78,	Arg. **2**, VI* 20°, Arg. **3**, III* 10°,	9,7	8,5
80,	IX* Équat. 15°,	22. 32,2	22. 33,2
85,	III* Diff. 3° to 5°,	5 6	6 7
88,	Arg. 1,1400 to 1,1500, Diff.,	3 2	30,2
92,	Conjonctions Moyennes, 1769,	5 0. 9.. 33,8	5 0. 9. 43,8
93,	Arg. **9**, 1772,	4. 3,2	4. 13,2
96,	Arg. **3**, 1863,	3. 23,33	2. 23,33
97,	Arg. **8**, 1857,	7. 28,8	2. 28,8
97,	Arg. II, 1857,	3. 16,96	3. 15,96
98,	Arg. **5**, 1867,	8. 23,8	4. 23,8
98,	Arg. **5**, Mai 1,	8. 2	8. 22,1
98,	Arg. **5**, Mai 8,	3. 2,1	3. 18,2
99,	Arg. IV. 1877,	8. 4,3	8. 5,3
100,	Arg. **5**. Mai 16,	14″,2	14°,2
109,	Diff. 1860,6 to 1860,7,	10,1	10,0
111,	VIII* 16°,	0,1	0,3
111,	IX* 16°,	0,3	0,1
115,	first column,	0° 4 3 2 1 5	0° 1 2 3 4 5
118,	Arg. **1**, V* 20°, Arg. **3**, X* 0°,	1,6	1,4
119,	Arg. **1**, VI* 10°, Arg. **3**, II* 0°,	1,6	1,4
119,	Arg. **1**, VIII* 10°, Art. **3**, 0* 20°,	2,0	2,1
119,	Arg. **1**, XI* 0°, Arg. **3**, V* 20°,	0,5	0,6
121,	Arg. **2**, VI* 20°, Arg. **3**, III* 10°,	9,7	8,5
122,	III* Équat. 11°,	4. 5,4	4. 5,5
122,	III* Diff. 10° to 12°,	0,5 0,4	0,4 0,5
123,	X* Équat. 6°,	0. 21,1	0. 21,0
123,	VII* Équat. 15°,	0. 37,7	0. 36,7
124,	top of columns 0° — XI*,	27°,9, 8°,7, *etc.*	27′,9, 8″,7, *etc.*
125,	first column,	0° 4 3 2 1 5	0° 1 2 3 4 5
125,	0* Équation 1°,	4. 12,2	4. 14,2
126,	VIII* Équation 11°,	8. 23,3	8. 23,1
127,	IV* Diff. 27° to 28°.	1,1	0,9
127,	VIII* Diff. 0° to 1°,	0″,7	0″,5
129,	V* Réduct. 3°,	2. 41,8	2. 41,0
133,	VI* Nombre 8°,	0,01028	0,01023
134,	V* 20°,	0,02547	0,00547
135,	VIII* 16°,	7,00352	0,00352
135,	VIII* 17°,	0,0x352	0,00352
136,	Arg. 0,16000, Demi-durées,	1. 6. 31,2	1. 6. 51,2
136,	second column of Arg.,	0,3100, 0,3200, 0,3300, *etc.*	0,31000, 0,32000, 0,33000, *etc.*
136,	Arg. 0,46000, N,	0,62	0,61

Page.		for	read
138,	top of page,	Suite de la TABLE XVIII	Suite de la TABLE XXVIII
146,	Arg. 3, 1840 B,	1. 26,92	1. 26,62

		for	read
149,	Arguments I and II, 1866 to 1880 B,	10. 24,50 2. 13,50 11. 25,07 3. 14,76 0. 25,66 4. 16,05 1. 26,30 5. 17,39 2. 25,59 6. 17,34 3. 26,24 7. 18,74 4. 26,94 8. 20,13 5. 27,65 9. 21,50 6. 28,28 10. 22,83 7. 27,49 11. 22,71 8. 28,06 0. 24,01 9. 28,62 1. 25,24 10. 29,18 2. 26,50 11. 29,76 3. 27,77 0. 28,99 4. 27,67	10. 25,00 2. 20,30 11. 25,57 3. 21,56 0. 26,16 4. 22,85 1. 26,79 5. 24,18 2. 26,06 6. 24,12 3. 26,76 7. 25,51 4. 27,45 8. 26,90 5. 28,12 9. 28,28 6. 28,77 10. 29,02 7. 27,99 11. 29,50 8. 28,57 1. 0,79 9. 29,14 2. 2,05 10. 29,70 3. 3,31 0. 0,28 4. 4,59 0. 29,50 5. 4,48

		for	read
149,	Arg. III, 1878,	2. 18,5	2. 16,5
164,	first column,	0° 4 3 2 1 5	0° 1 2 3 4 5
166,	II* Diff. 10° to 11°,	7,6	7,5
167,	Arg. 1, I* 20°, Arg. 3, X* 0°,	3,1	3,8
167,	Arg. 1, IV* 20°, Arg. 3, VII* 10°,	3,3	3,1
167,	Arg. 1, IV* 20°, Arg. 3, VII* 20°,	3,3	2,9
170,	Arg. 2, VI* 20°, Arg. 3, III* 10°,	9,7	8,5
171,	VIII* 19°,	6,6	5,6
177,	III* 20°,	2,57582	2,57562
180,	first column,	0° 4 3 2 1 5	0° 1 2 3 4 5
181,	Arg. 0,50000 to 0,51000, Diff.,	1976	1970
183,	II* Diff. 27° to 28°,	25	26
183,	IV* Diff. 10° to 11°,	20	19
184,	second column of Arg. Q + Z,	0,02250	0,00250
187,	Arg. Q + Z, 0,64000 to 0,65000, Diff.,	53,8	53,3
(190),	X* Équation 27°,	+11. 5 ,7	+11. 57,7
(193),	eleventh line,	l'élongation est entre 3* et 0*	l'élongation est entre 9* et 0*
(216),	Longit. moy., 1849,	4. 5,32	0. 4,53

		for	read
(224),	Argument 4, 1870 to 1880 B,	4. 23,0 3. 5,7 1. 18,5 0. 22,8 11. 5,5 9. 18,3 8. 1,0 7. 5,3 5. 18,1 4. 0,8 2. 13,6	4. 23,8 3. 6,6 1. 19,3 0. 23,7 11. 6,4 9. 19,1 8. 1,9 7. 6,2 5. 18,9 4. 1,7 2. 14,4